Chi-wai Kan

Ka-yan YIM

Resin Treatment of Light-weight Plain Cotton Knitted Fabric

AF138161

Chi-wai Kan
Ka-yan YIM

Resin Treatment of Light-weight Plain Cotton Knitted Fabric

LAP LAMBERT Academic Publishing

Impressum / Imprint
Bibliografische Information der Deutschen Nationalbibliothek: Die Deutsche Nationalbibliothek verzeichnet diese Publikation in der Deutschen Nationalbibliografie; detaillierte bibliografische Daten sind im Internet über http://dnb.d-nb.de abrufbar.
Alle in diesem Buch genannten Marken und Produktnamen unterliegen warenzeichen-, marken- oder patentrechtlichem Schutz bzw. sind Warenzeichen oder eingetragene Warenzeichen der jeweiligen Inhaber. Die Wiedergabe von Marken, Produktnamen, Gebrauchsnamen, Handelsnamen, Warenbezeichnungen u.s.w. in diesem Werk berechtigt auch ohne besondere Kennzeichnung nicht zu der Annahme, dass solche Namen im Sinne der Warenzeichen- und Markenschutzgesetzgebung als frei zu betrachten wären und daher von jedermann benutzt werden dürften.

Bibliographic information published by the Deutsche Nationalbibliothek: The Deutsche Nationalbibliothek lists this publication in the Deutsche Nationalbibliografie; detailed bibliographic data are available in the Internet at http://dnb.d-nb.de.
Any brand names and product names mentioned in this book are subject to trademark, brand or patent protection and are trademarks or registered trademarks of their respective holders. The use of brand names, product names, common names, trade names, product descriptions etc. even without a particular marking in this works is in no way to be construed to mean that such names may be regarded as unrestricted in respect of trademark and brand protection legislation and could thus be used by anyone.

Coverbild / Cover image: www.ingimage.com

Verlag / Publisher:
LAP LAMBERT Academic Publishing
ist ein Imprint der / is a trademark of
OmniScriptum GmbH & Co. KG
Heinrich-Böcking-Str. 6-8, 66121 Saarbrücken, Deutschland / Germany
Email: info@lap-publishing.com

Herstellung: siehe letzte Seite /
Printed at: see last page
ISBN: 978-3-659-53192-7

Zugl. / Approved by: The Hong Kong Polytechnic University, BA Thesis, 2011

Resin Treatment of Light-weight Plain Cotton Knitted Fabric

Chi-wai Kan and Ka-yan Yim

Institute of Textiles and Clothing,

The Hong Kong Polytechnic University,

Hung Hom, Kowloon, Hong Kong

Abstract

The purpose of this study was to investigate the wrinkle resistant and dimensional stability of light weight plain cotton knitted fabrics. As cotton knitted fabric has the tendency to wrinkle and its structure is instable during finishing process and customer usage, wrinkle free finishing is needed to improve the properties of cotton knitted fabric.

In this study, experimental investigation was conducted to assess the wrinkle resistant performance and the dimensional properties of plain knitted fabrics. Four types of plain knitted fabrics with various weight, yarn court, courses and wales and cover factor were treated with varied resin concentration and were subjected to five home laundering cycles. The effectiveness of resin treatments on wrinkle resistant and dimensional stability of each type of fabric was analyzed based on induced wrinkle and the dimensional change on the fabric surface. Furthermore, research was conducted to investigate the air permeability and bursting strength of the knitted fabrics.

Comparison of cotton fabrics treated with resin finishing is discussed. General conclusions and recommendations have been done about the overall performance of cotton fabric.

CONTENT

	Page No.
ABSTRACT	**2**
CONTENT	**3**

CHAPTER 1	**INTRODUCTION**	
1.1	Background of Study	7
1.2	Objectives	8
1.3	Scope of Study	9
1.4	Research Methodology	9
1.5	Significance of Study	10
1.6	Arrangement of Thesis	10

CHAPTER 2	**LITERATURE REVIEW**	
2.1	Introduction	12
2.2	Cotton	12
2.2.1	Cotton Fiber	12
2.2.2	Chemical Properties of Cotton	13
2.2.3	Physical Properties of Cotton	15
2.3	Knitted Fabric	17
2.3.1	Plain Knit Structure	17
2.3.2	Fabric Specification	18
2.3.3	Dimensional Stability	21
2.3.3.1	Types of Dimensional Changes	23
2.3.3.2	Standard Test Method	23
2.4	Resin Treatment	24
2.4.1	Wrinkle	24
2.4.2	Resin Treatment	24

	2.4.3	Development of Resin Treatment	25
	2.4.4	Application of Resin Treatment	26
	2.5	Conclusion	26

CHAPTER 3 **METHODOLOGY**

	3.1	Introduction	27
	3.2	Fabric Preparation	27
	3.2.1	Fabric Details	27
	3.2.2	Pre-treatment	28
	3.2.3	Fabric Specification	28
	3.3	Resin Treatment	29
	3.3.1	Introduction	29
	3.3.2	Apparatus	29
	3.3.3	Experimental Procedure	30
	3.4	Dimensional Stability to Repeated Home Laundering	30
	3.4.1	Introduction	30
	3.4.2	Sample Preparation	31
	3.4.3	Apparatus	31
	3.4.4	Experimental Procedure	32
	3.4.5	Calculation and Interpretation	32
	3.5	Smoothness Appearance of Fabrics after Repeated Home Laundering	33
	3.6	Wrinkle Recovery of Fabrics	33
	3.6.1	Introduction	33
	3.6.2	Sample Preparation	34
	3.6.3	Apparatus	34
	3.6.4	Experimental Procedure	35

3.6.5   Evaluation   36

3.7   Air Permeability   37

3.7.1   Introduction   37

3.7.2   Experimental Procedures   38

3.8   Bursting Strength   38

3.8.1   Introduction   38

3.8.2   Experimental Procedures   39

3.9   Conclusion   40

CHAPTER 4   WRINKLE RESISTANT PERFORMANCE AND DIMENSION PROPERTIES OF COTTON KNITTED FABRIC

4.1   Introduction   41

4.2   Smoothness Appearance   41

4.2.1   Result   42

4.2.2   Discussion   44

4.3   Dimensional Stability   46

4.3.1   Result   47

4.3.2   Discussion   53

4.4   Wrinkle Recovery   54

4.4.1   Result   58

4.4.2   Discussion   59

4.5   Conclusion   59

CHAPTER 5   PHYSICAL PROPERTIES OF COTTON KNITTED FABRICS

5.1   Introduction   60

5.2   Air Permeability   60

	5.2.1	Introduction	**60**
	5.2.2	Result	**61**
	5.2.3	Discussion	**62**
	5.3	Bursting Strength	**63**
	5.3.1	Introduction	**63**
	5.3.2	Result	**64**
	5.3.3	Discussion	**66**
	5.4	Conclusion	**68**

CHAPTER 6 CONCLUSION AND RECOMMENDATIONS

	6.1	General Conclusion	**70**
	6.2	Recommendation	**70**

REFERENCE **73**

Chapter 1 Introduction

1.1 Background of Study

Being a popular natural cellulosic fiber, the worldwide consumption of cotton has growing million metric tons annually. Although the cotton prices in global economy are rapidly increasing, the versatility, durability and comfort of cotton are among the factors for it remaining a preferred state and major textile material in apparel industries.

However, the major demerit of cotton knitted fabric is the tendency to wrinkling and dimensional changes upon different condition and this is the targeted problem to be discussed in my study. Due to the fact that the demand for easy care clothing and wrinkle free cotton product has raised and "no-iron" becomes an expectation of consumers, the added functional value of wrinkle resistant can strengthen the competitiveness of cotton knitted fabric in the textile industry.

Over the year, the morphology, chemistry and utilization of cotton have been discussed by many researchers. Many cotton fabrics are treated with chemicals to reduce wrinkling. The wear resistance of cotton fabric is a complex phenomenon which depends on fabric weight, fabric geometry, yarn structure and size, and fiber properties (Gagliardi & Wehner, 1967). Refurbishing like laundering and dry-cleaning will increase the tendency of wrinkle cotton fabric. To improve the wrinkle-resistant properties of cotton fabric, resin finishing by pad-dry-cure method is applied

7

in this project.

In addition, the state of wrinkling becomes severe on knitted fabric, and the dimensional stability of knitted fabric is varying while laundering or wearing. Knitted fabric is characterized by its elastic properties. A fair amount of deformation in the shape of the fabric is possible when external forces are applied. In theory, the fabric would recover its natural shape and dimensions. The dimensional behavior of knitted fabric depends on the particular physical characteristics of the fiber in the yarn which are balanced by the geometry of interlocking loops (Lo et al., 2009). However, different yarn court and density of knitted fabric will affect its performance under different condition. These criterions influence the quality of fabric and can be improved depend on the ability of recovery from induced wrinkles and dimensional changes. Dimensional change is the change in lengthwise or widthwise of a fabric subjected to different conditions. The change is usually expressed as a percentage of the original dimension of the fabric. Resin treatment makes cotton knitted fabric more stable in its dimension during washing process. In this project, cotton knitted fabric treated with varied resin concentration was subjected to one, three and five home laundering cycles. The effectiveness of resin treatments on dimensional stability was investigated based on dimensional change in fabric form.

1.2 Objectives

The principle objectives of this study are summarized as follows:

1. To evaluate the effectiveness of varied concentrations of resins on controlling wrinkle resistant and dimensional stability of cotton knitted fabrics

2. To investigate the effect of home laundering on fabrics wrinkling and dimensional changes on cotton knitted fabrics

3. To compare the wrinkle resistant and dimensional stability of resin treated cotton fabrics with different yarn court and cover factor

4. To assess the performance of cotton knitted fabric by test methods such as smoothness appearance, wrinkle recovery, air permeability and bursting strength

1.3 Scope of Study

The scope of this study is to evaluate the performance of resin treated cotton knitted fabrics in terms of wrinkle resistant, dimensional stability, air permeability and bursting strength. Relationships between these properties to the plain knitted fabric are discussed.

1.4 Research Methodology

In order to achieve the objectives, the following methodologies have been adopted:

1. Literature reviews will be conducted in order to acquire the background knowledge and recent development in the relevant areas

2. Easy care finishing will be applied to cotton knitted fabrics as wrinkle resistant treatment, using various concentrations of crosslinking solution by pad-dry-cure

method

3. International standards such as AATCC, ASTM and ISO test methods will be adopted for evaluation and comparison.

4. Evaluations on the wrinkle resistant performance and dimension properties will be performed and further investigation on physical properties of knitted fabrics will be conducted

1.5 Significance of Study

This study aims at improving the wrinkle-resistant performance and dimensional stability of light cotton knitted fabric. The importance of the application of resin treatment to enhance these properties will be evidenced. Furthermore, the relationship of resin treatment level and wrinkle resistant and dimensional changes was investigated after the experimental details.

1.6 Arrangement of the book

This book comprises of six chapters.

In Chapter 1, the background, objectives, scopes, methodology and significance of this research will be introduced.

In Chapter 2, relevant literature reviews will be conducted, mainly concerning the cotton fiber and molecular, the structure of cotton knitted fabric and the development and application of wrinkle resistant technology in the textile industry.

Chapter 3 covers the cotton knitted fabrics details and specifications, and experiment details including apparatus used, sample preparation and procedures based on the resin treatment, AATCC, ASTM and ISO test methods to evaluate wrinkle resistant and dimensional stability of knitted fabrics during various wash and wear condition.

In Chapter 4, evaluation and interpretation will be performed to assess the effect of resin treatment and repeated home laundering on cotton knitted fabrics. The capability of wrinkle resistant and dimensional stability is discussed using test result and graphs.

In Chapter 5, the physical properties of cotton knitted fabrics are conducted to investigate the air permeability and bursting strength of the knitted fabrics.

Chapter 6 is the general conclusion of this study. Recommendations are also provides for further study.

Chapter 2 Literature Review

2.1 Introduction

Due to the challenge of synthetic fiber like polyester, the demand of cotton fabrics is shifting from the performance of aesthetic, versatility and durability, to the properties of wrinkle-free and easy care. Wash and wear garments are preferred by consumers.

To fulfill such demand appeared in the market, wrinkle resistant finishing is needed to enhance the properties of the fabrics. Information about the materials and techniques applied in this study are summarized in this chapter. With the aim of exploring the effect of resin treatment on cotton fabrics and making comparison, this chapter also reviews relevant knowledge of cotton fiber, knitted fabric and resin treatment.

2.2 Cotton

2.2.1 Cotton Fiber

Cotton is the most common type of fiber in the textile industry and is currently the leading fiber worldwide. Cotton is also the most important textile fiber, as well as cellulosic textile fiber; about 38% of the fiber consumed is cotton. Cotton fibers are natural cellulosic fiber from the boll of cotton plant. Its kidney-bean, cross-sectioned shape, convolution along its length, and the hydroxyl group on the cellulose polymer contribute to the performances of cotton fibers. Those characteristics of cotton fiber enhance its performances due to the increase in moisture regain, absorption rate, smoothness of surfaces, and comfort sensation to skin.

As cotton is a highly crystalline fiber, the weak strength of covalent bond and stronger of hydrogen bond allows the cotton fibers in having medium strength, low elongation at break, low abrasion resistance, and low elasticity and elastic recovery. The convolution of fiber shape gives next-to-skin comfort. The hydrophilic property of cotton absorbs moisture from skin. The origin, development, morphology, chemistry, purification and utilization of cotton have been discussed by many authors. An overall review has been done on the physical and chemical properties of cotton.

2.2.2 Chemical Properties of Cotton

Chemical Structure

Cotton is a natural cellulosic fiber which its chemical composition consists of 95% cellulosic. Cotton contains carbon, hydrogen, and oxygen with reactive hydroxyl (-OH) groups. Molecular chains of glucose are arranged in long spiral linear chains within the fiber. The strength of cotton fiber is related to the cellulosic chains length.

From the molecular structure of cotton, several hydroxyl (-OH) groups outside the circle are composed of one oxygen and five carbon atoms. Hydrogen bonding occurs between adjacent cellulosic chains in crystralline areas of fiber when the hydrogen atoms of the hydroxyl (-OH) group are attracted to the oxygen atoms. The bonding of hydrogen's within the ordered regions of the fibrils causes the molecules to draw closer to each other which increases the strength of the fiber. Cotton consists about 70% crystallinity and 30% amorphous region, but the crystalline regions are not oriented as the electrons circumjacent the atoms are not evenly scattered (Project Cotton, 2008). The element composition of cotton is: cellulose (95%); protein (1.3%); ash (1.2%); wax (0.6%); sugar (0.3%); organic acids (0.8%) and other chemical compounds (3.1%) (Project Cotton, 2008).

Moisture Absorption

Hydrogen bonding is an important role for the properties of cotton fiber. It remedies in moisture absorption. Hydrogen bonds that contribute to the comfort of cotton make cotton ranking among the most absorbent fibers. The hydrophilic property of cotton fiber results in a high water absorption. But the hydrogen bonds will be easily broken and wrinkle or creases are formed easily on the surface of fabric due to the broken hydrogen bonds. Thus, 100% cotton fabric easily wrinkles and is potential to shrink after laundering. In resin-treated cotton, the hydroxyl groups on adjacent fibrils are permanently joined by the covalent groups of the resin, so as to provide wrinkle resistance to the fabric.

Chemical Reaction

Cotton is affected and reacts to chemicals in many different ways. Cotton is very sensitive to acid since it is hydrolyzed and forms hydrocellulose when in contact with acids. In contrast, cotton has excellent resistant to alkalis, making itself swelling.

2.2.3 Physical Properties

Morphology

From the physical point of view, cotton molecule is a ribbon-like structure of linked with hydroxyl groups. The covalently bonded chain molecule is stiffened by internal hydrogen bonds. The molecule has high modulus and high strength and it has high rigidity for bending in the plane.

Each cotton fiber has a single plant cell. It consists of four major components, which are a cuticle, primary wall, secondary wall and a lumen. The cuticle is a few molecules thick and covers the primary wall with a waxy film. Primary wall is formed by the growth to its full length and diameter. It consists of numerous fibrils spiraling around the fiber axis. Fibrils are simply packs of cellulose chains. The secondary wall is formed in daily growth rings inside the tube, leaving a small lumen at the centre when the cotton is mature. The secondary

15

wall has several layers of spiraling fibers, which make up most of the weight of the cotton fiber. The lumen, located within the secondary wall, is a hollow canal that carries nutrients of the cotton during growth (Parker, 1998).

A mature cotton fiber is u-shaped and like kidney-bean (Hearle, 2007). Convolutions of cotton are natural ribbon-like twists along the length of the fiber. When cotton fiber matures, lumen will dry out and collapses which makes secondary wall start to twist.

Strength

High strength and extensibility of cotton make it as a good textile fiber. Cotton has a breaking tenacity of 3.5 to 4.0 grams per denier, making it fairly strong (Project Cotton, 2008). As cotton has adequate degree of crystallinity, the strength of the fiber is moderate. When cotton is wet, it becomes about 20% stronger. In terms of elasticity, cotton is poor. It breaks after moderate stretching and does not recover well. As a result, cotton tends to stay stretched in areas of stress like knees and elbows. The strength of cotton fabric can be evaluated as bursting strength.

Comfort

Cotton has many characteristics that help to account for its comfort. It is very soft that feeling comfortable against the skin. Cotton is also good in warm and humid weather because it has good ventilation and superior absorbency, which moisture can be able to freely pass through the fabric. Cotton also has good heat and electrical conductivity that keeps static from being a nuisance. (Kadolph, 2010)

2.3 Knitted Fabric

2.3.1 Plain Knit Structure

Knitted fabrics are defined as an ensemble of loops bonded together in an elastic mode. A knitted fabric is composed of a yarn which is bent into repeated loops across the width and along the length of a fabric. Plain knit fabrics have the main popularity in the marketplace. Plain knit fabric is composed of knitted stitches and made from just one needle bed only. Utilizing one single bed to knit means the plain knit show all face loops on one side and all back loops on another side. The fabric is said to have different appearances and unbalanced structures is resulted. The yarn can be unraveled from the course broken.

Due to the high extensibility in both length and width, knitted fabrics allow garments to fit closely and snugly, making them ideal for next-to-skin wear. Thus, knitwear is becoming more and more popular in fashion industry recently especially cotton ones due to its good moisture absorption leading to superior comfort.

However, knitted fabrics are prone to stretching and mechanical deformations. This is due to the fact that the yarns are put under a high stress factor whilst the fabric is being produced and finished. These torsion forces within the yarns are present when the fabric is taken off the machine and the fabric is left in a highly distorted state (Anand, et al., 2002). As a result, the fabric shrinkage has been a problem for maintaining the quality of the fabrics. In the common practice of our fashion industry, controlling the dimensional change of knitted fabrics has been the main concern of manufacturers to satisfy their customers' need.

Light cotton knitted fabrics has a higher tendency of fabric shrinkage and low wrinkle resistance. Cotton is a cellulose material which has poor resistance to wrinkling. The levels of shrinkage and distortion occurred in cotton knits increase substantially under repeated home laundering and tumble drying.

2.3.2 Fabric Specification

Fabric weight

Fabric weight can be measured in grams per square meter. It determines the thickness of a fabric. Fabric weight of knitted fabric is difficult to measure precisely due to the fiber loss while cutting.

Yarn court

It expresses the thickness of the yarn, and must be known before calculating the quantity of yarns for a known length of fabric. The yarn count number indicates the length of yarn in relation to the weight. The Tex system is based on metric weights and measures. Tex is an internationally agreed system of yarn numbering that applies to all types of yarns, regardless of the method of production (Knitting-and, 2010). Yarn court is equal to linear density. Variations in yarn diameter affect the performances of fabric. It is expressed as mass per unit length or length per unit mess.

$$yarn\ court\ (tex) = \frac{W_T}{L_T} \times 1000$$

Where WT = total weight (in mg) and LT = total length (in mm).

Course and wale

They are the repeated units and ridges created in the knitted fabric (shown in Fig 2-1). Courses are the ridges that run widthwise in the fabric; wales run lengthwise. The number of courses and wales (per inch) which contain within the area of the fabric sample is counted. The measurement is called course and wale density. Stitch density is counted by course density and wale density. It gives a total number of stitches in a square area of a fabric. Stitch density tends to give more accurate measurement for fabric dimensions (Ucar, 2007).

The construction of a fabric today is still frequently described in terms of courses and wales per unit length. The use of this approximate and averaged

19

parameter-for specifying the tightness of a knitted construction is responsible (directly or indirectly) for many of the problems associated with the dimensional control of knitted structure (Abou-ana, et al., 2003).

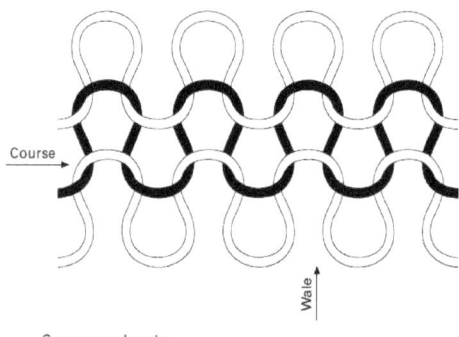

Course and wale.

Fig 2-1 Course and wale of knitted fabric

Loop Length

The length of yarn knitted into a single loop is very important and it determines such overall fabric qualities as hand, comfort, weight, extensibility, finished size, cover factor, and most importantly, fabric dimensional stability (Munden, 1959).

Cover Factor

Cover factor indicates the extension which the region of a knitted fabric is covered or formed by its yarn. It is also an indication of relative looseness or tightness of knitted fabric structure. Cover factor is also called tightness factor and its unit is K. Yarn is removed from specific knitted loops and its length is measured under sufficient tension to remove the knitting crimp. The known length of yarn is weighed to obtain the yarn count. Cover factor is then calculated from the yarn court and the yarn length per knit loop. From the theoretical point of view, the loop length can be calculated as

$$L = \frac{\text{yarn length}}{100} \, ;$$

The cover factor for plain knitted structures as

$$K = \frac{\sqrt{\text{tex}}}{L}$$

Where, L = loop length, 100 = 100 loops, K= cover factor, and tex = yarn court.

2.3.3 Dimensional Stability

Dimensional stability indicates a fabric's ability to resist changes in its dimension. Dimensional control is important apparently in plain knitted fabrics. A fabric or garment my contain shrinkage, increase or decrease in dimensions, under different condition or refurbishing. In the view of knitting, structural parameters such as length of the loop and the number of stitches in wale and course would give some information about knitted fabrics' dimensional stability.

In the after-care practices of households, the main activities involve machine laundering and drying. The washing process is a cooperative interaction of both physical and chemical effects to the garment. Home laundering influences the dimensional stability of the fabric under repeated washing cycle. Fabrics are especially affected by the water temperature used in washing and drying temperature in tumble drying.

During weaving of fabrics, it is subjected to considerable tensions, after subsequences of finishing such as crosslinking, the stretch is set in the fabric temporarily. When the fabric is wetted, it tends to revert to its more stable dimensions so lead to dimensional instability after washing.

The dimensional stability of knitted fabric is poor without any finishing treatment. The natural reaction of yarn in a knitted state was to attempt to return to its previous straight state, and because this could not be achieved, it would take up a state in which the total energy in the fabric was at a minimum (Nutting, 1960). Actually, it is very rare to find a yarn unraveled from a knitted fabric to be perfectly straight. On the other hand, during knitting, the yarn is bent into loops with larger curvature, especially near the needle loop and sinker loop (Choi & Lo, 2003). The dimensional stability of cotton fabric can be enhanced by the finishing of resin.

2.3.3.1 Type of Dimensional Changes

The dimensional behavior of fabrics depends on the pattern of the fabric and the fiber type used. Plain knit fabrics have more dimensional changes in the coursewise direction than the walewise direction. The most common dimensional change of a fabric is shrinkage. In the length and width direction of a fabric can be shrinking at different percentage. This often more pronounced in knitted fabrics. Some instance is growth in the width while shrink in length.

2.3.3.2 Standard Test Method

Dimensional stability test is designated to see how well a fabric keeps its shape after washing and drying. It is essential for knitted fabric to be tested. The traditional method for testing fabrics is the measurement of length and width benchmarks before and after laundering process.

There are many test methods to examine the dimension properties of fabrics. In this study, AATCC Standard Test method 135: Dimensional Change of Fabrics after Home Laundering was used. This test method is suitable for knitted fabric. Appropriate benchmarks are applied and test specimens are placed in an automatic washing machine with ballast added to make a full wash load. Evaluation on dimensional changes of test specimens can be done after laundering cycle.

2.4 Resin Treatment

2.4.1 Wrinkle

Wrinkles are defined as the fabric deformations based on its viscoelastic properties mean a slight depression in the smoothness of a surface. Cotton wrinkles easily when worn, it shrinks after first wash then gradually from every wash after and it does not recover well from stretching without any pretreatment or finishing. Therefore, modification of wrinkle-resistant of cotton fabric is needed to reinforce the mechanical performance.

2.4.2 Resin Treatment

The history of wrinkle resistant treatment can be traced back to 1920's when the research scientists at Toolal Broadhurst Lee Company started trying wrinkle-free effect on cellulosic fibers. The term of wrinkle resistant and crease resistant were defined as easy care or wash-and-wear in the 1950's. The need for ironing after home laundering has eliminated due to the effect of durable press and resin finish.

The wrinkling behavior of cotton fabric is related to the free hydroxyl groups in the amorphous regions, which are bound to each other. To impart wrinkle resistant finish to the cotton material, the hydrogen bond formation of the hydroxyl groups should be either masked or totally removed. Wrinkle resistant finishes aims to form covalent bonds crosslinking between cellulose chains that

24

replace the weaker hydrogen bonds and provide greater stability in the position of molecules.

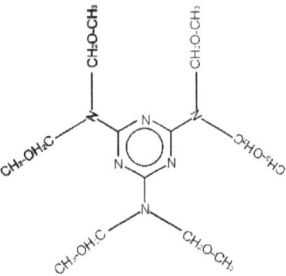

Fig. 2-2 Resin Crosslinking

2.4.3 Development of resin treatment

Treatment with a crosslinking agent, urea formaldehyde DMDHEU, was used to enhance the fabric properties and prevent winkles and creases in the early stage of wrinkle free development. But it had been proved that such agent damaging cotton fabric, and weakening the abrasion resistant, tear strength, etc. Formaldehyde resin has been restricted due to the fact of environmental harming and being a hazardous substance. Substitutions have been developed as low formaldehyde and non-formaldehyde finishing agents. Modified glyoxal-based reactants decrease the formaldehyde release to a tiny fraction. Polycarboxylic acid derivatives produce fabrics with better abrasion resistant, moderate durable press performance and lower shrinkage.

2.4.4 Application of resin treatment

As concerning the loss in tensile strength of cotton due to depolymerization of cellulose chains, crosslinker for the low-formaldehyde easy care finishing was used in this study. Fixapret F-ECO is a modified dimethyloldihydroxyethylene urea that provides very good stability to hydrolysis and soft handle. With magnesium chloride as acid catalyst to initiate the reaction, a neutral pH can be obtained in aqueous solution in all temperatures and offer maximum bath stability.

2.5 Conclusion

The background and physical and chemical properties of cotton material is focused in this chapter. Further background knowledge of plain knit structure and resin treatment is also obtained. The details of the experimental procedures of the rein finishing and dimension test will be demonstrated in Chapter 3.

Chapter 3 Methodology

3.1 Introduction

In this chapter, experiments on cotton knitted fabrics will be described; showing the details of apparatus and materials used, preparation of test specimens, testing procedures, calculation and interpretation.

Four types of cotton knitted fabrics were treated with various concentration of low-formaldehyde crosslinking solution by pad-dry-cure method, in order to see how these parameters influenced the properties of wrinkle resistant and dimensional stability of the treated fabrics.

Further evaluations were conducted so as to investigate the smoothness appearance, dimensional changes and wrinkle recovery of cotton knitted fabric after subjecting to repeated home laundering test.

3.2 Fabric Preparation

3.2.1 Fabric Details

Commercially produced light-weight knitted fabrics from combed cotton with four different yarn's linear densities were used in the study. Each fabric lot was bleached and treated with fluorescent brightening agent (FBA). The gray fabric weight of type A, B, C and D is 100gsm, 110gsm, 160gsm and 120gsm respectively. The yarn court of each fabric is 11.81tex, 14.76tex, 29.53tex and 18.45tex.

3.2.2 Pre-treatment

Before the resin treatment, cotton fabrics were scoured to wash away any impurity. Non-ionic detergent was used with 1g/L concentration. The total weight of Fabric A, B and D is the same of 2.1 kg and Fabric C weigh 2.4 kg. The liquor-to-goods ratios are 34:1 and 29:1 respectively. The total weight of liquor employed in the treatment is 70 liters.

Laboratory pad-stream was used for the treatment. Cotton fabrics were scoured at room temperature. Raised the temperature to 90°C and boiled fabrics for 10 minutes. Then rinsed the fabrics three to four times before dried flat in oven.

3.2.3 Fabric Specification

The scoured fabrics were cut into the size of 380 mm x 380 mm. The Tables 1 to 3 below show the fabric specification of each type of fabric after scoured.

Table 3-1 Fabric weight and yarn court of cotton knitted fabrics

Fabric Type	Fabric Weight (gsm)	Fabric Yarn Court (tex)
A	107.1	13.85
B	130.2	18.46
C	197.4	32.31
D	144.9	21.54

gsm = grams per square meter ($\pm$ 1)

Table 3-2 Course and wale density per inch

Fabric Type	Course Density	Wale Density
A	46	53
B	42	50
C	31	37
D	39	46

Density per in.: ±1

Table 3-3 Loop length and cover factor

Fabric Type	Loop Length (cm)	Cover Factor (K)
A	0.27	13.78
B	0.29	14.82
C	0.38	14.96
D	0.32	14.50

Equation: $$K = \frac{\sqrt{tex}}{L}$$

Where: Loop length (L) ±0.01 cm

3.3 Resin Treatment

3.3.1 Introduction

Resin treatment of low formaldehyde finishing was applied on four types of cotton knitted fabrics after scoured. It aims to impart wrinkle resistant of cotton fabric in order to enhance the stability of fabric shape.

3.3.2 Apparatus

Resin treatment of cotton knitted fabrics was carried out by pad-dry-cure method.

3.3.3 Experimental Procedure

Three different concentration of resin solution was applied to the fabric specimens using a horizontal padder with a 70% pick-up. 500ml solution was prepared for padding (Table 3-4).

Different parameters including speed, pressure and number of time of padding were set depending on the fabric properties. After padding, the specimens were dried in the oven at 70°C for 5 minutes, and then cured for 1 minute at 170-175°C. Table 3-5 shows the various padder parameters for different fabric types.

Table 3-4 Resin concentrations used

Recipe	1	2	3
Fixapret F-ECO	30g/L	40g/L	50g/L
Magnesium Chloride	8g/L	9g/L	10g/L

Table 3-5 Horizontal padder parameters used

Fabric Type	Speed (RPM)	Pressure	Padding time
A	5	2	2
B	5	2	3
C	3.5	1.4	3
D	3.5	1.4	2

RPM: revolutions per minute

3.4 Dimensional Stability to Repeated Home Laundering

3.4.1 Introduction

To evaluate the dimensional changes of fabrics when subjected to repeated home laundering, specimens were washed for five cycles according to the standard test method AATCC 135-2004. Machine cycle, washing temperature and drying condition were assigned for a completed cycle of laundering.

3.4.2 Sample Preparation

380 mm x 380 mm resin treated specimens were used. Laying the fabric on a flat surface, six pairs of benchmarks parallel to the test specimen length and width were then marked. Each benchmark was 50 mm from all edges of the specimens (Fig 3-1).

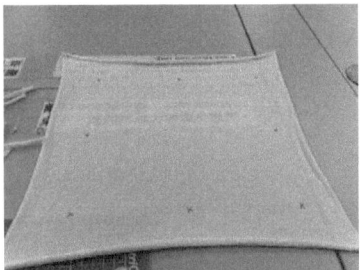

Fig 3-1 Fabric preparation of dimensional stability

3.4.3 Apparatus

Automatic washing machine and automatic tumble dryer were used (Fig 3-2).

Fig 3-2 Automatic washing machine (left) and tumble dryer (right) used

3.4.4 Experimental Procedure

Washing

1. Weigh test specimens and enough ballast to make a 1.8 ± 0.1 kg load

2. Select medium water level, $27 \pm 3\,^{\circ}\text{C}$ water temperature and cold rinse temperature for the washing cycle

3. Add 66.0 ± 1 g of washing solvent to the wash load

4. Set the machine cycle to permanent press

5. Start the machine until the cycle end signal

Tumble Drying

1. Place the washed test specimens and ballast in the tumble dryer

2. Set the temperature as permanent press - high exhaust temperature

3. Allow the dryer operate until the total load is dry

3.4.5 Calculation and Interpretation

Test specimens were taken out for dimension measurement every after 1st, 3rd and 5th wash-and-tumble drying cycle. Fabrics were laid without tension on a flat smooth surface. The distance between each pair of benchmarks were measured and recorded. The percentage of dimensional changes on average length and width of test specimens were calculated using equation as follows:

Average% DC = 100 (B – A)/A

Where: DC = average dimensional change

A = average original dimension

B = average dimension after laundering

The result will be discussed on Chapter 4.

3.5 Smoothness Appearance of Fabrics after Repeated Home Laundering

After repeated home laundering, the smoothness appearance of test specimens was evaluated according to AATCC 124-2009. It aims to assess the effect of washing to the performance of wrinkle resistant of cotton knitted fabrics. Fig. 3-3 shows two test specimens on the viewing board. Fabric smoothness grades were rated in the specific room. Fluorescent light was the only light source for observation to the board. Fabrics surfaces were observed and adjusted to various level of fabric smoothness or freedom from wrinkles.

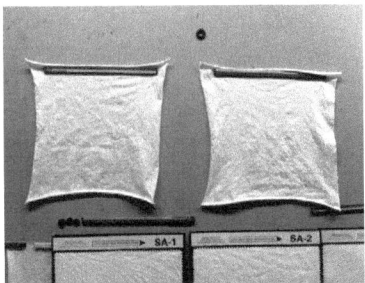

Fig 3-3 Two test specimens on the viewing board and compare with smoothness appearance replica equivalents

3.6 Wrinkle Recovery of Fabrics

3.6.1 Introduction

AATCC Test Method 128-2009 Wrinkle Recovery of Fabrics: Appearance Method was simulated to evaluate the surface of knitted fabrics after induced wrinkle. It aims to assess how severe of crease effect on the elbow and knee area of garment. Test specimens were reconditioned for evaluation.

3.6.2 Sample Preparation

Test specimens were cut into the size of 6 x 11 in. with the long dimension running in the direction of the wale of cotton knitted fabrics (Fig 3-4).

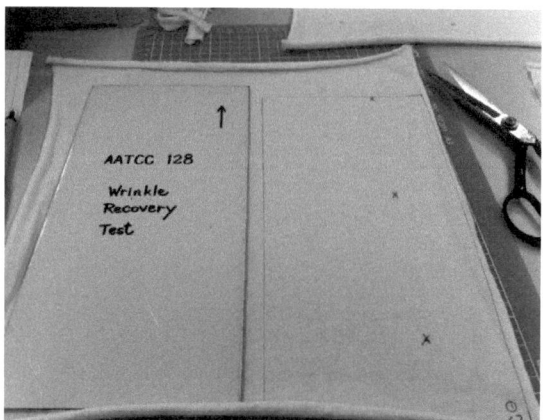

Fig 3-4 Sample preparation of wrinkle recovery test

3.6.3 Apparatus

Fig 3-5 shows the AATCC Wrinkle Tester used in this test. Addition 3700 grams weight set was place on the top flange.

Fig 3-5 AATCC Wrinkle tester

3.6.4 Experimental Procedure

Top flange of the wrinkle tester was hold in top position with locking pin. Test specimens were wrapped, facing outwards, around the upper and lower flanges of the tester (Fig 3-5). Steel springs were used to clamp the specimens to the flange. Withdrawing the locking pin to lower the top flange and placing the weight on it can thereby induce wrinkling (Fig 3-6). After 20 minutes, the weight should immediately remove from the top flange (Fig 3-7). Raised the flange and springs and clamps were removed to take away the specimens from the tester (Fig 3-8).

Fig 3-6 Wrinkle recovery test

Fig 3-7 After wrinkle recovery test

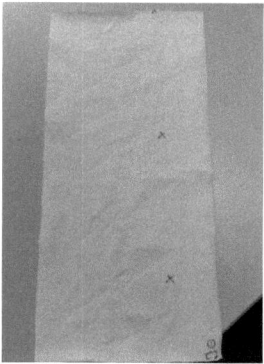

Fig 3-8 Test specimen with induced wrinkle

3.6.5 Evaluation

Two evaluations were taken; one was right after the test and another was 24 hours later. Test specimens were mounted on the viewing board after the first observation and waited 24 hours for the second observation (Fig 3-9). Both evaluations took place at the darkened room and ratings were given according to the wrinkle recovery replicas (Fig 3-10).

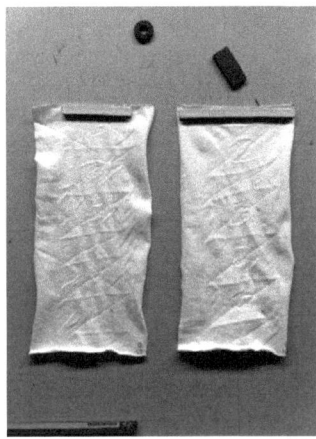

Fig 3-9 Evaluation on wrinkle recovery of cotton knitted fabric

Fig 3-10 Wrinkle recovery replicas

3.7 Air Permeability

3.7.1 Introduction

Air permeability can be used to indicate the breathability of treated fabrics. As knitted fabrics have high porosity, they have relatively better air permeability among other fabrics. To estimate the efficiency of air permeability of the resin treated fabrics, ASTM D737 Standard Test Method for Air Permeability of Textile Fabrics was used.

3.7.2 Experimental Procedures

Air permeability testing apparatus was used to draw a steady flow of air perpendicularly through the test specimens (Fig 3-11). Air flow rate and water pressure can be adjusted manually according to the fabrics being tested.

In this test, the air flow rate and water pressure should be adjusted for those four types of cotton knitted fabrics. Each test specimen was placed on to the test head of the test instrument and water pressure differential was read and recorded.

Fig 3-11 Air permeability apparatus

3.8 Bursting Strength

3.8.1 Introduction

The purpose of bursting strength test is to measure the fabric strength in all direction at the same time and therefore is suitable for knitted fabrics. The bursting strength test was conducted with accordance of the standard test method ISO 13938-2 Bursting Properties of Fabrics: Pneumatic method for determination of bursting strength and bursting distension. Fig 3-12 shows the instrument of the test – bursting tester.

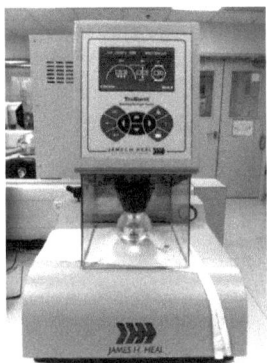

Fig 3-12 Bursting Tester

3.8.2 Experimental Procedures

Test specimens were placed over the elastic diaphragm and laid in a flat condition. A circular holder was used to clamp the specimen in order to avoid slippage. Tests were performed to the area of specimens within the circular clamping device. Pressure rate of the bursting tester can be adjusted automatically or manually, depends on the fabrics properties. After started the bursting tester, increasing compressed air pressure was applied to the underside of the diaphragm. Distension of the diaphragm and the specimens caused by the pressure was increased smoothly until the fabric burst (Hunter 2007).

Bursting pressure and height at burst were noted. After repeated the test five times at different area on the fabric, five sets of data were collected. Diaphragm correction was needed before the calculation so as to distend the diaphragm without the presence of test specimens. The arithmetic mean of the bursting pressure (psi: pound per square inch), the height at burst values (mm) and time of rupture (s) were calculated. Standard deviation and percentage of coefficient of variation were also recorded.

3.9 Conclusion

This experimental work is focus on the study of examining wrinkle resistant and dimensional stability of cotton knitted fabrics treated with resin finishing and the effect of repeated home laundering. The physical properties of air permeability and bursting strength of the knitted fabrics were also examined. In this chapter, resin treatment and home laundering test method are discussed to explain the experimental details of the research work. Further discussion and investigation on the performance of wrinkle-free and dimensional changes is illustrated in the next chapter.

Chapter 4 Wrinkle Resistant Performance and Dimension Properties of Cotton Knitted Fabrics

4.1 Introduction

The previous chapter discussed about how wrinkle-resistant treatment was applied to cotton knitted fabric. Tests such as repeated home laundering and wrinkle recovery were also described to explain the experimental details.

In this chapter, the result of wrinkle resistant performance including smoothness appearance and wrinkle recovery, and the percentage of dimensional changes are investigate and evaluated in order to compare the effect of various levels of resin treatment and reiterated home laundering on light-weight plain cotton knitted fabrics. A deeper study on the properties of wrinkle resistance and dimensional stability can be generated.

4.2 Smoothness Appearance

After treated with resin crosslinking solution by pad-dry-cure method, four type of cotton knitted fabrics (Fabric A, B, C, and D) have undergone five wash-and-tumble drying cycles so as to simulate the effect of domestic laundering. Evaluations were performed every after 1st, 3rd and 5th cycle of washing, using standard lighting and viewing room, rating the visual appearance of specimens quantified by comparison with the set of reference standard replicas from AATCC 124-2009.

4.2.1 Result

AATCC Test Method 124 is designed for evaluating the appearance, in terms of smoothness, of flat fabric specimens after repeated home laundering. This provides a measure of the durable-press and easy-care, or minimum iron properties of the fabric. The test procedure and evaluation method about specimen preparation and standard replicas has discussed in the chapter above. Four figures were generated to assess the effect of resin treatment to the smoothness appearance of the fabrics.

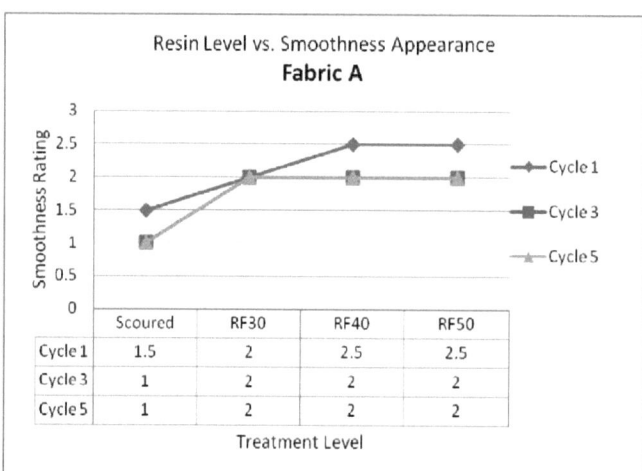

Fig. 4-1 Relationship between resin level and the smoothness appearance of Fabric A

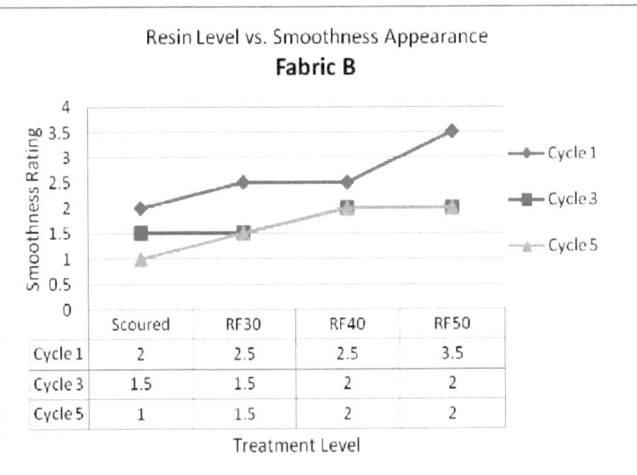

Fig. 4-2 Relationship between resin level and the smoothness appearance of Fabric B

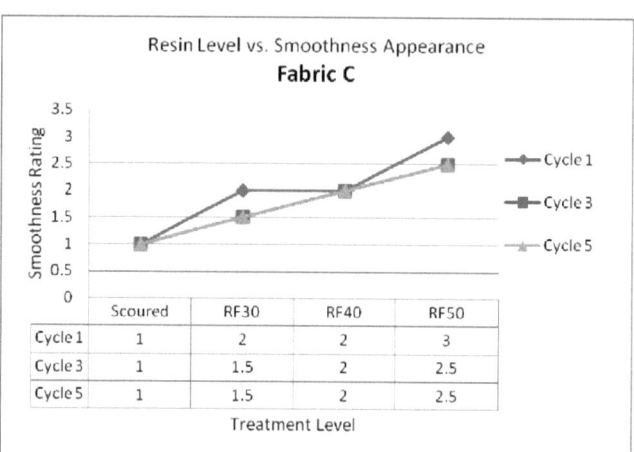

Fig. 4-3 Relationship between resin level and the smoothness appearance of Fabric C

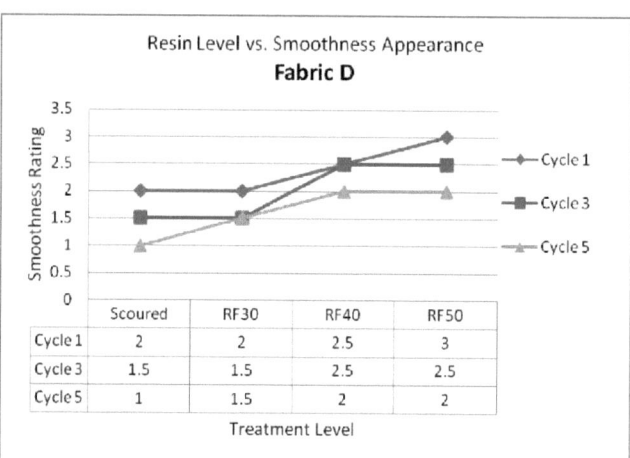

Fig. 4-4 Relationship between resin level and the smoothness appearance of Fabric D

4.2.2 Discussion

Set of samples for observation was prepared for each of the four different knitted fabrics. In each set, there were a scoured sample and three samples with different levels of resin treatment. Evaluation results were recorded after the repeated home laundering test. Line charts are used to describe the tendency of smoothness of each type of fabric.

From the above Figures 4-1 to 4-4, there are linear relationships between the resin treatment level and smoothness appearance rating. Number of washing and tumble drying cycle also has obvious influence on the fabric.

Effect of resin treatment on the smoothness surface of cotton fabrics

In term of treatment level, it is believed that the level of resin finishing did affect the performance of smoothness appearance of cotton knitted fabric. From Figure 4-4, the ratings of smoothness of fabrics increase while the resin finish level grows. There is a trend showing that the fabrics have obviously wrinkled appearance when less resin concentration was used.

The higher the resin level, the fairly smoother the fabric surface is resulted. It indicates that the concentration of resin treatment has a direct effect on the wrinkle property of the fabrics. This also proved that the higher resin level can effectively enhance the fabric wrinkling property and has efficiency impact on the fabric surface.

Effect of wash-and-tumble drying cycle on the smoothness appearance of fabric

Other than the resin treatment factor, the number of washing and tumble drying cycle also affect the impression of planarity of the fabrics. From Figures

4-1 to 4-4, all fabric samples have undergone five washing cycle and they clearly show that, when more laundering is applied to the fabrics, the wrinkle effect will more severe than that with less cycle.

Take Figure 4-2 as an example, the smoothness rating of Fabric B is decreasing when the number of laundering cycle increase. Three separate lines are resulted on the figure, stating that the rating of cycle 1 is relatively higher than cycle 3 and 5. It directly shows the effect of repeated home laundering on the surface property of knitted fabric.

4.3 Dimensional Stability

Excessive changes in fabric dimensions can represent a serious problem in virtually all textile applications, more particularly in clothing. Thus, dimensional stability to laundering forms an important quality and test requirement. Dimensional change has been defined as a generic term for percentage changes in the length or width of a fabric specimen subjected to specific conditions. Standardized washing machines and tests, to assess fabric or garment performance under repeated home laundering cycles, have been developed as AATCC 135, which includes both washing and tumble drying. The actual changes in fabric dimensions during a test depend upon a number of factors, such as yarn court, fabric weight, stitch density and tightness factor.

4.3.1 Result

An industry norm for dimensional stability test on cotton single jersey is, for example, no more than 8% shrinkage in either length (wale) or width (course) direction (Hunter, 2007). Figures 4-5 to 4-18 outlines the dimensional change results. The relationship between the dimension property and the resin treatment will be discussed.

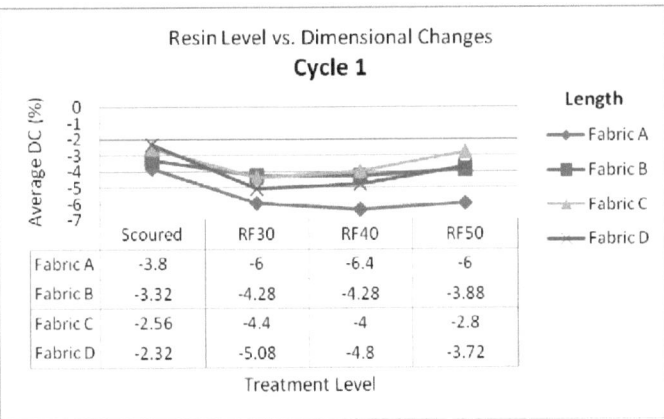

Fig. 4-5 Dimensional changes of wash-and-tumble drying Cycle 1 (Length)

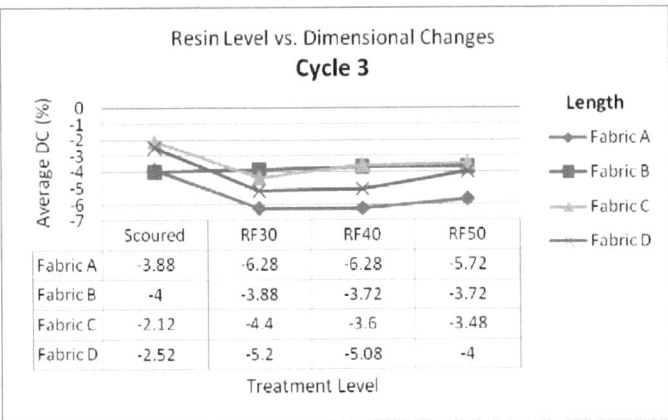

Fig. 4-6 Dimensional changes of wash-and-tumble drying Cycle 3 (Length)

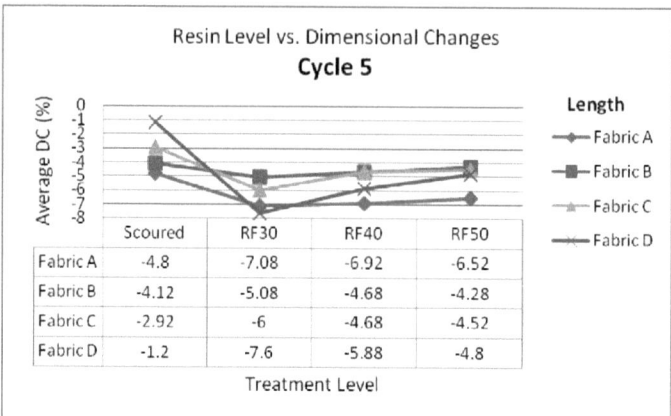

Fig. 4-7 Dimensional changes of wash-and-tumble drying Cycle 5 (Length)

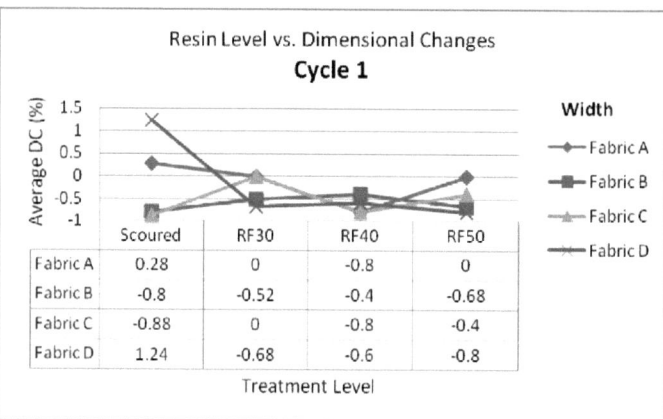

Fig. 4-8 Dimensional changes of wash-and-tumble drying Cycle 1 (Width)

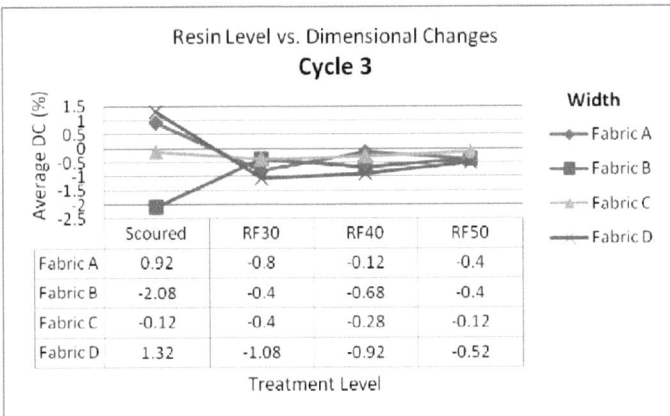

Fig. 4-9 Dimensional changes of wash-and-tumble drying Cycle 3 (Width)

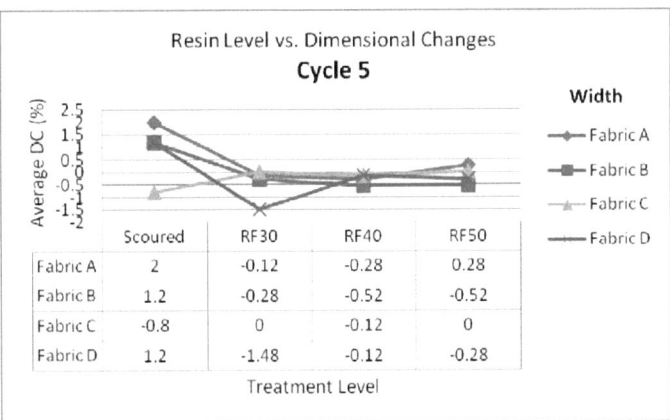

Fig. 4-10 Dimensional changes of wash-and-tumble drying Cycle 5 (Width)

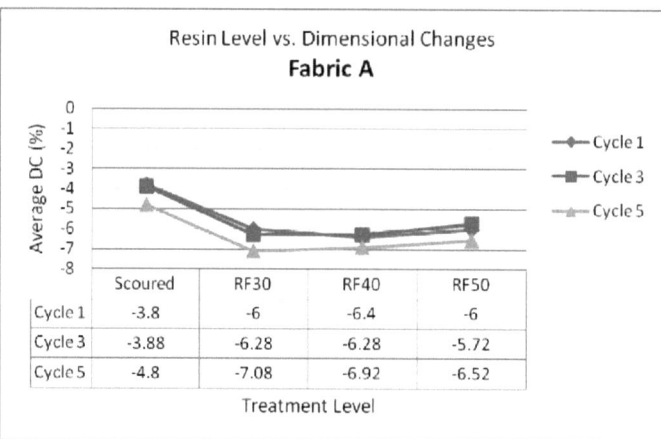

Fig. 4-11 Relationship between resin level and the dimensional changes of Fabric A (Length)

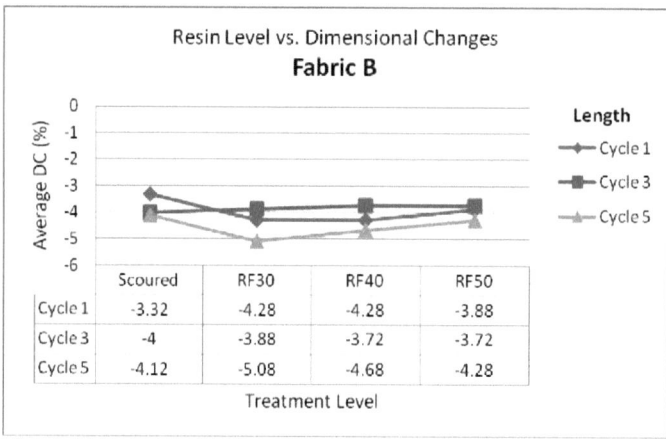

Fig. 4-12 Relationship between resin level and the dimensional changes of Fabric B (Length)

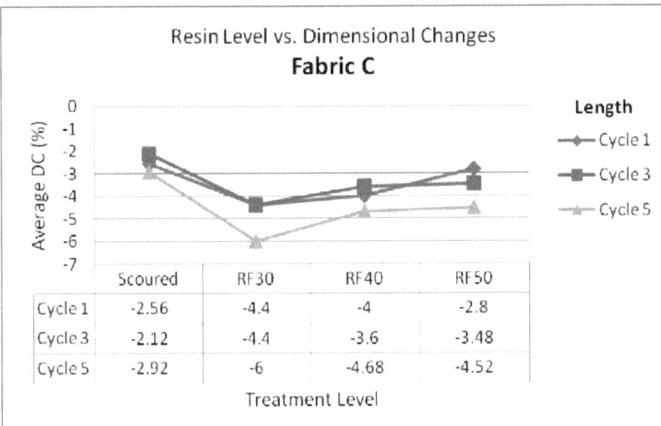

Fig. 4-13 Relationship between resin level and the dimensional changes of Fabric C (Length)

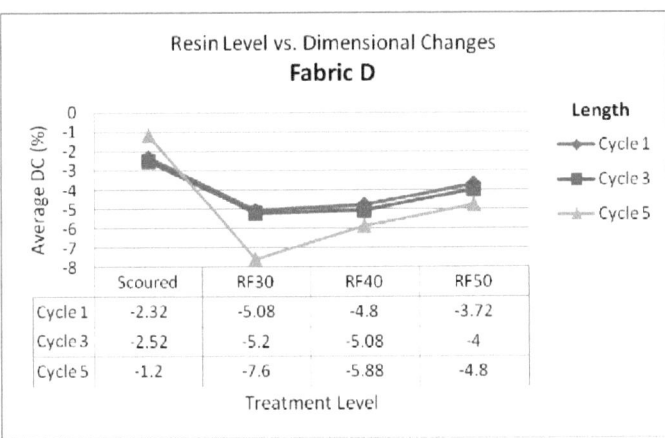

Fig. 4-14 Relationship between resin level and the dimensional changes of Fabric D (Length)

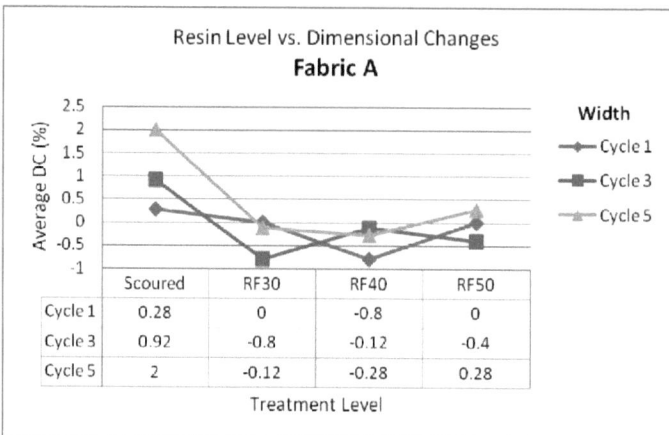

Fig. 4-15 Relationship between resin level and the dimensional changes of Fabric A (Width)

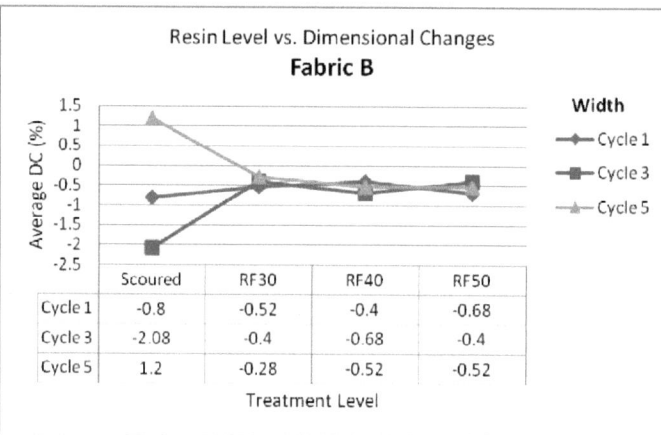

Fig. 4-16 Relationship between resin level and the dimensional changes of Fabric B (Width)

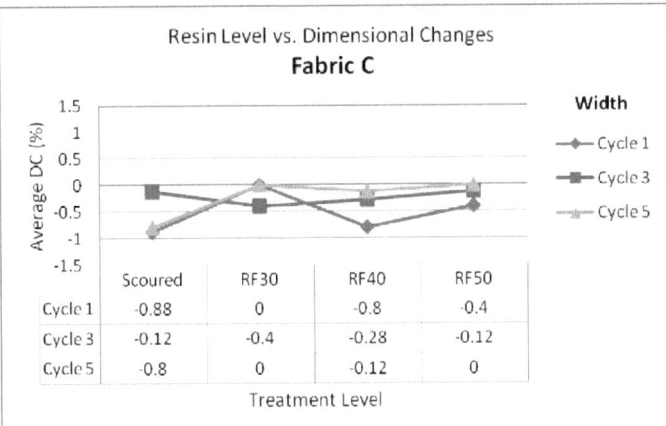

Fig. 4-17 Relationship between resin level and the dimensional changes of Fabric C (Width)

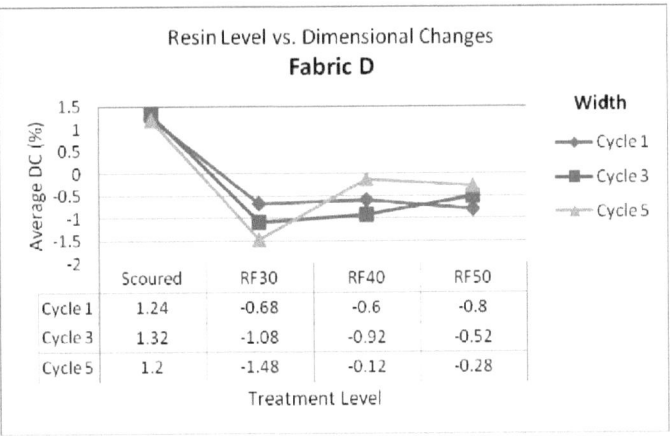

Fig. 4-18 Relationship between resin level and the dimensional changes of Fabric D (Width)

4.3.2 Discussion

From Figures 4-5 to 4-18, the percentage of dimensional changes in length and width direction of cotton fabrics has been calculated. As expected, the percentage change in length direction is larger than the width direction.

In addition, the effect of repeated home laundering has shown on the graphs. When the washing and drying cycle increase, the knitted fabrics tend to have a larger percentage of dimensional changes. It indicates that increased laundering condition change the shape of fabric more severely. Shrinkage problem is more serious while having much more time of washes.

However, there is no clear relationship between the resin treatment level and the dimensional stability of cotton fabric. The fluctuation of data is more severe in width direction of the fabric. As a result, it can be said that there is no linear relationship.

4.4 Wrinkle Recovery

The tendency for a fabric or garment to wrinkle during wear or laundering, is very important from an appearance point of view, also having a bearing on 'ease-of-care' related properties. In effect, it is not so much the ability of the fabric to withstand wrinkling, but its ability to regain its original shape and smooth appearance and its wrinkle recovery is important.

Therefore, wrinkle recovery is particularly important for cotton, since untreated cotton is notoriously poor in this respect and considerable research and development work over many decades has been directed towards developing chemical treatments that improve this property without an unacceptable loss in other desirable properties.

A popular method used by industry to assess the fabric wrinkle recovery is

AATCC Test Method 128 Wrinkle Recovery of Fabrics: Appearance Method in which wrinkles are induced in the fabric under standard atmospheric conditions using a standard wrinkling device under a predetermined load for a prescribed period of time. The specimen is then reconditioned and rated for appearance by comparing it with three-dimensional reference standards.

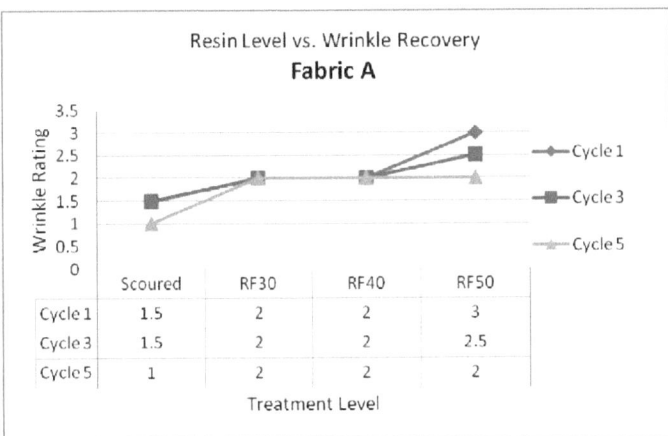

Fig. 4-19 Relationship between resin level and the wrinkle recovery of Fabric A

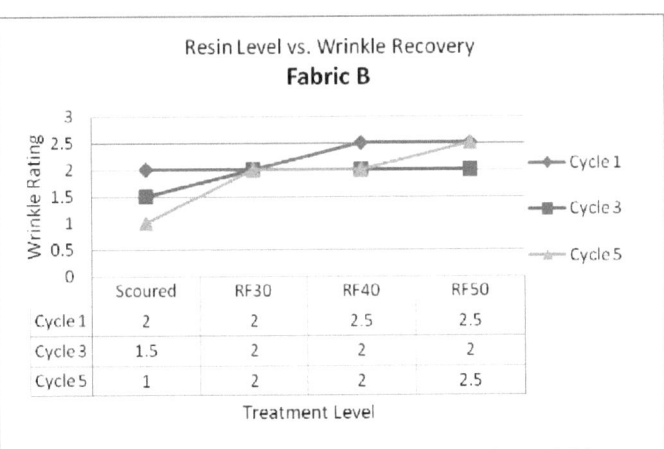

Fig. 4-20 Relationship between resin level and the wrinkle recovery of Fabric B

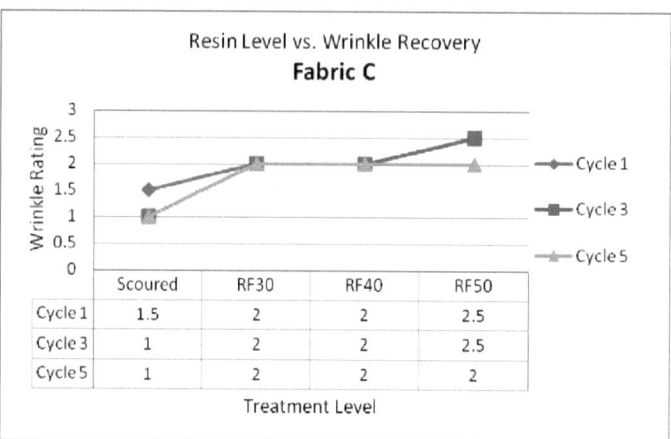

Fig. 4-21 Relationship between resin level and the wrinkle recovery of Fabric C

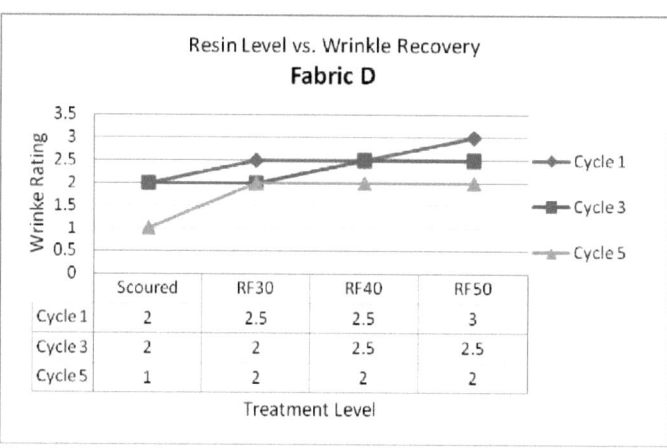

Fig. 4-22 Relationship between resin level and the wrinkle recovery of Fabric D

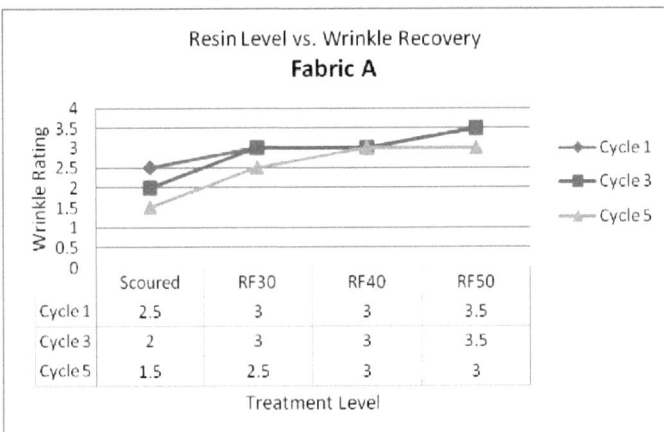

Fig. 4-23 Relationship between resin level and wrinkle recovery of Fabric A after 24h

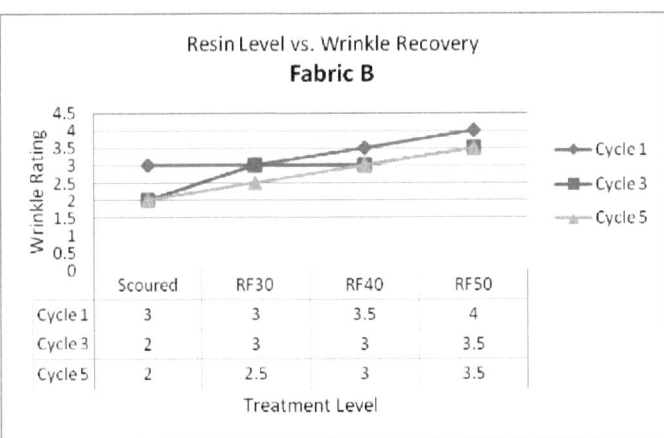

Fig. 4-24 Relationship between resin level and wrinkle recovery of Fabric B after 24h

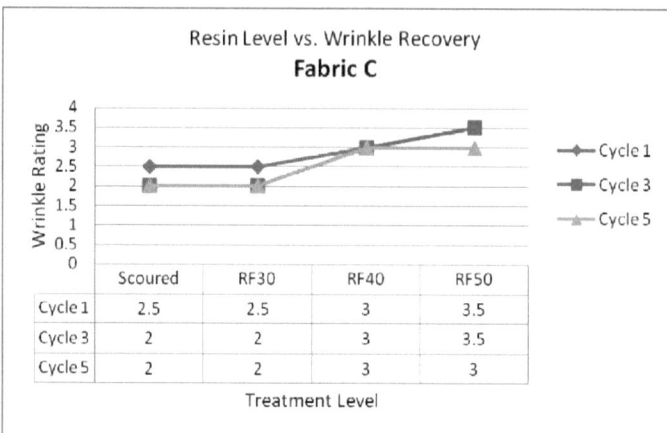

Fig. 4-25 Relationship between resin level and wrinkle recovery of Fabric C after 24h

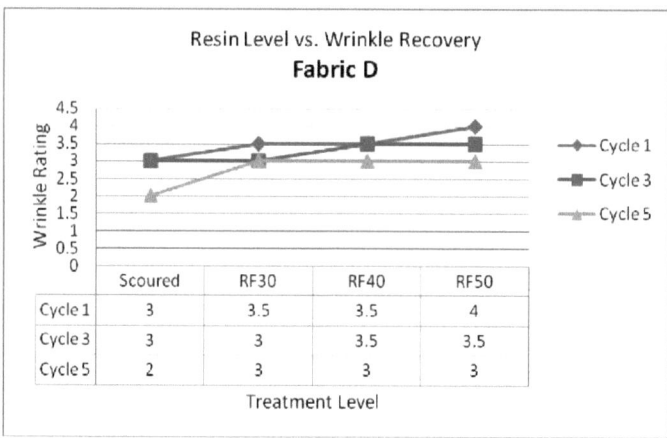

Fig. 4-26 Relationship between resin level and wrinkle recovery of Fabric D after 24h

4.4.1 Result

Two evaluations were conducted after the wrinkle recovery test. One is right after the test and the other is 24h later. It aims to investigate how effectiveness the fabric recovers to its original shape and surface. Wrinkle performance is assessed by the above Figs 4-19 to 4-26.

4.4.2 Discussion

The wrinkle resistant is believed to be related to the level of resin treatment. The resin concentrations reflect the tendency of wrinkling in the cotton knitted fabrics. It has shown in Figs 4-19 to 4-22 that a higher level of resin treatment resulted in a higher rating of wrinkle recovery.

Besides, the knitted fabrics have a higher rating after 24 hours. Fair retention of original appearance and smooth appearance is obtained. Figs 4-23 to 4-26 show that the fabric recovery evenly to its original surface. A higher rating of wrinkle recovery is recorded with the effect of resin treatment.

4.5 Conclusion

In this Chapter, the relationship between smoothness appearance and resin level was studied and the results revealed that the level of resin finishing did affect the smoothness appearance of the cotton knitted fabrics. In addition, the number of laundering cycle also affected the surface appearance of resin-treated cotton knitted fabrics. Meanwhile, when the washing and drying cycles were increased, the resin-treated cotton knitted fabrics tended to have a larger percentage of dimensional change.

Chapter 5 Physical Properties of Cotton Knitted Fabrics

5.1 Introduction

The chapters above have mentioned about the performance of wrinkle resistant and dimension properties of cotton knitted fabrics. But the special porosity structure of knitted fabrics and the tensile strength of cotton can also affect the performance of garments under different wearing condition.

In this chapter, the physical properties of resin treated cotton knitted fabrics would be investigated. Several physical evaluation tests were performed. Air penetrability can be assessed by using the air permeability test. The fabric strength would also be measured according to ISO standard test method.

5.2 Air Permeability

5.2.1 Introduction

Due to its porosity property, most cotton knitted fabrics have better air penetrability than others. But the degree of air permeability of knitted fabrics can be affected by different fabric specifications. Diverse cotton court, stitch density and tightness factor influence the air permeability performance of such textile materials. In order to measure the air permeability of resin treated cotton knitted fabrics, the following test were designated.

5.2.2 Result

Result of the air permeability is determined in accordance with ASTM Test Method D737. Four types of knitted fabrics were tested and results were grouped into 3 tables (Figs 5-1 to 5-3) by continuous washing and tumble drying cycles: 1st, 3rd and 5th.

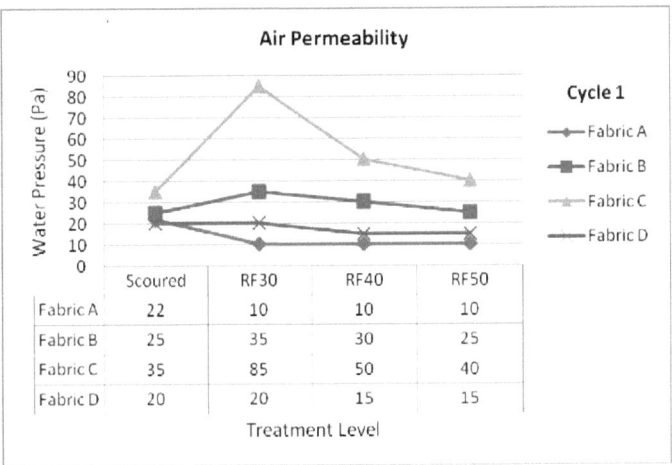

	Scoured	RF30	RF40	RF50
Fabric A	22	10	10	10
Fabric B	25	35	30	25
Fabric C	35	85	50	40
Fabric D	20	20	15	15

Fig 5-1 Air permeability of wash-and-tumble drying Cycle 1

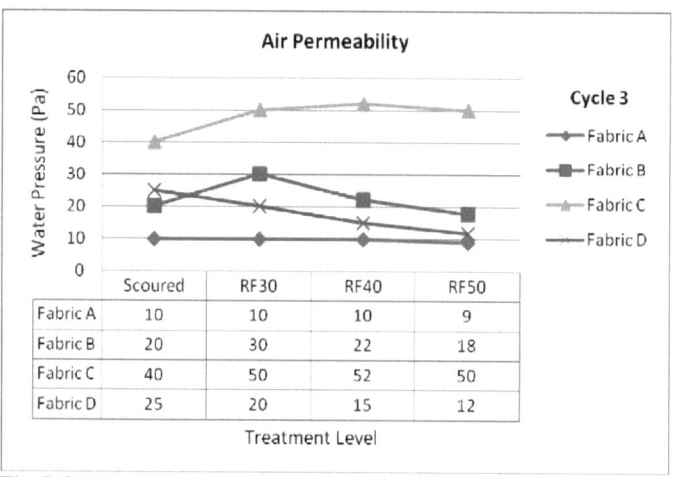

Fig 5-2 Air permeability of wash-and-tumble drying Cycle 2

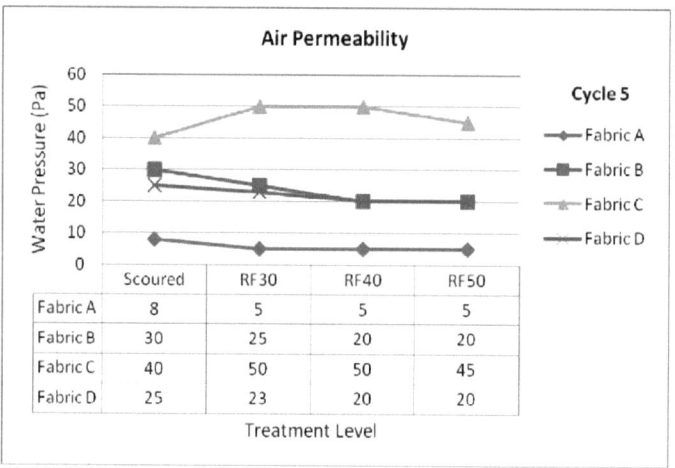

Fig 5-3 Air permeability of wash-and-tumble drying Cycle 3

5.2.3 Discussion

Figs 5-1 to 5-3 show the results in the air permeability of plain cotton knitted

fabrics in different resin treatment applied on the fabrics. From the above Figs 1 to 3, we can observe that there is no apparent effect of different resin finishing on the performance of air permeability in cotton fabric. It can be seen that no obvious difference is appear when the level of resin treatment increased. The water pressure required of each fabric sample remains unchanged from scoured to RF50 treatment level.

On the other hand, a close relationship between the air permeability and fabric properties has shown in the figures. From the experimental results, different fabrics contain different cover factor. Generally speaking, the larger the cover factor, the lower the air permeability is resulted.

5.3 Bursting Strength

5.3.1 Introduction

The efficiency of bursting strength required for a fabric mainly depends on its end use. For example, industrial garments need a relatively higher strength of fabrics. Garments intended for apparel and household use require adequate strength so as to withstand various wearing condition. For some cases, higher bursting strength implies stiffer and heavier fabrics with undesirable handle and stiffness.

Bursting strength aims to measure fabric strength in all direction at the same time and is therefore suitable for knitted fabrics. Elastic diaphragm is used to load the fabrics. There is pressure of fluid behind the diaphragm to measure the stress in the fabrics. Additional pressure or force is needed to inflate the diaphragm

5.3.3 Result

Four types of cotton fabrics treated with resin finishing and repeated home laundering were tested in dry condition by bursting strength test. Comparisons of bursting strength of the fabrics are investigated in terms of resin treatment level and wash-and-tumble drying cycle.

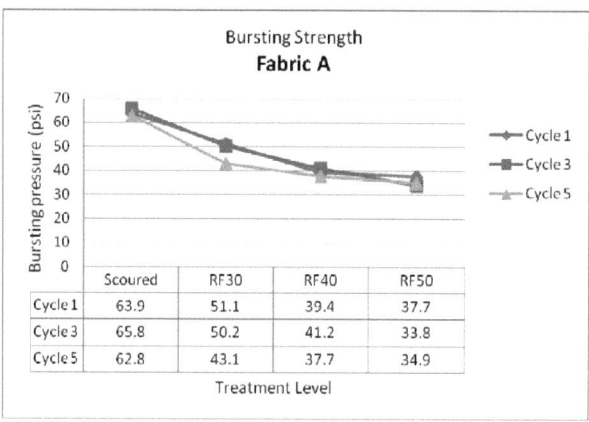

Fig 5-4 Bursting Pressure of Fabric A

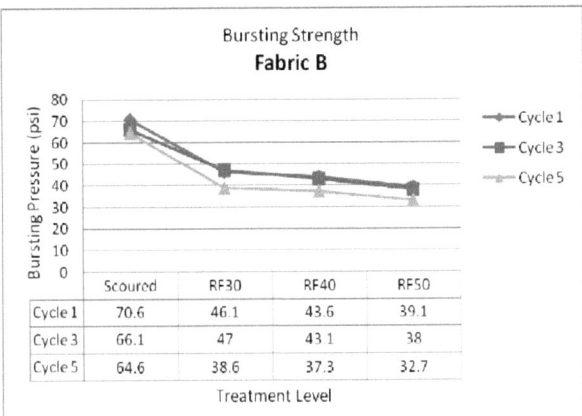

Fig 5-5 Bursting Pressure of Fabric B

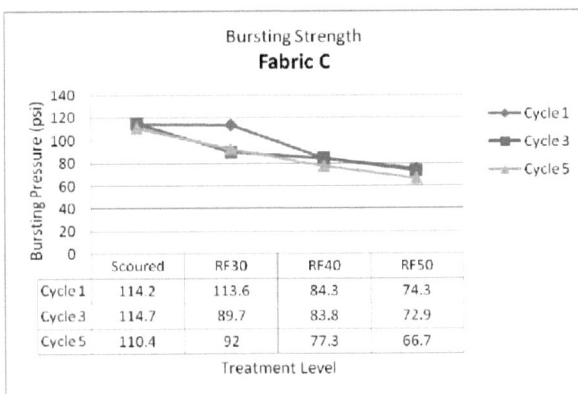

Fig 5-6 Bursting Pressure of Fabric C

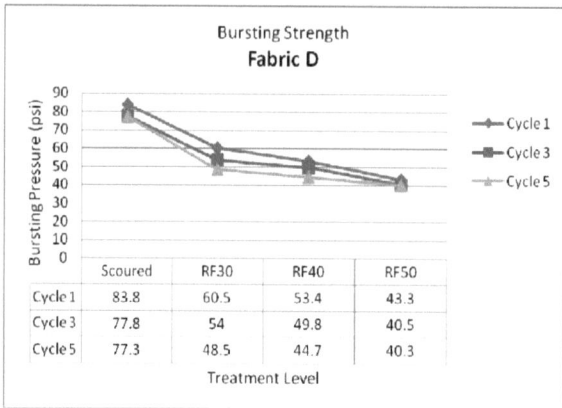

Treatment Level	Scoured	RF30	RF40	RF50
Cycle 1	83.8	60.5	53.4	43.3
Cycle 3	77.8	54	49.8	40.5
Cycle 5	77.3	48.5	44.7	40.3

Fig 5-7 Mean Bursting Pressure of Fabric D

5.3.4 Discussion

Physical and mechanical properties of cotton knitted fabrics can be affected by several fabric parameters. And bursting strength would be affected by fabric weight and cover factor. Resin treatment and laundering test can also influence the bursting strength of knitted fabric. Comparisons are made in terms of fabric parameters, various resin treatment level and different washing cycles.

Fabric Parameter

Air permeability of knitted fabrics depends on several factors such as thickness, fabric density, yarn properties, etc. As discussed in the chapter of experimental details, Fabric A, B, C and D are four different cotton knitted fabrics with various fabric specifications. In term of fabric weight, Fabric C is

66

the heaviest knitted fabric and Fabric A is the lightest one. At the same circumstance, cover factor of each fabric is also determined.

From Figs 5-4 to 5-7, different fabric burst at different bursting pressure. Fabric C requires the highest bursting pressure of 110 - 115 psi and Fabric A requires the least having 63 - 66 psi. Fabric B and D are in the middle bursting pressure of 67 - 70 psi and 76 - 84 psi respectively. This indicates that heavier fabric requires a relatively larger bursting pressure to rupture the fabric. The cover factor of fabrics generally exhibits less air permeability due to less free movement of the loops and space rotation within the fabric structure. The tighter the cover factor, the higher the bursting pressure is used to burst the fabric.

Resin Treatment Level

Applying different levels of resin treatment to cotton knitted fabrics, we can have different measure of bursting pressure. Figures 5-4 to 5-7 have been shown that the bursting strength of Fabric A, B, C and D is much related to the application of resin finish. The above figures confirm that there is a clear relationship between the level of resin treatment and the bursting strength of fabrics.

For each type of fabric, the bursting pressure (psi) was plotted against various states of resin treatment. Take Fabric B as example, a higher level of resin treatment resulted in a lower bursting pressure. RF 30 required 46.1 psi to rupture the fabric but RF 50 required less bursting pressure of 39.1 psi. As a result, it is expected that a higher level of resin finishing may weaken the fabric strength and less bursting pressure is required to burst the fabric.

Wash-and-tumble drying cycle

Not only the resin treatment level affecting the bursting strength of the fabrics, more wash-and-tumble drying cycles can also influence the bursting capability of plain knitted fabrics.

Figures 5-4 to 5-7 also reflect the effect of repeated home laundering to the cotton fabrics. It can be seen that, increased washing and tumble drying cycles with decrease in bursting strength of Fabric A, B, C and D. The more the laundering cycles, the less the bursting required rupturing the fabrics. Figure 5-7 shows that while Fabric D have undergone more washing cycles; there is a decrease in bursting pressure. The bursting strength of the fabric is reduced obviously. Therefore we can say that the bursting strength is believed to be related to the repeated home laundering.

5.4 Conclusion

In the previous content, the relationship between the level of resin treatment to the air permeability and bursting strength has been discussed. In term of air permeability, the crucial factor affecting the result is the cover factor of cotton knitted fabrics. It is believed that larger cover factor results in relatively lower air permeability. In term of bursting strength, both the resin treatment and washing cycle influence the capability of fabric strength. It is mainly because the resin finishing is directly weakening the bursting strength of cotton fabrics.

Chapter 6 Conclusion and Recommendations

6.1 General Conclusion

In this study, the influences of different fabric parameters, various resin concentration used in the resin treatment and the effect of repeated home laundering to the properties of plain cotton knitted fabrics have been explored. The results obtained lead to the following conclusion.

1. There were enhancements on the performance of cotton knitted fabric after resin treatment such as smoothness appearance, dimensional stability and wrinkle recovery.

2. There was no relationship between the resin treatment level and the air permeability, but it is affected by cover factor of the fabric.

3. There was a reduction on the bursting strength of cotton knitted fabric after resin treatment.

4. Higher concentration of resin finishing could effectively provide a moderate wrinkle resistant property on the fabrics.

6.2 Recommendations

In this highly competitive global market, the survival of a textile material will greatly depend upon its ability to meet demanding quality specifications through optimized manufacturing and testing. To strengthen and enhance the properties of fabrics, easy care finishing is highly recommended.

In the present work, the influences of different fabric parameters, various resin concentration used in the resin treatment and the effect of repeated home laundering to the properties of plain cotton knitted fabrics have been studied. However, further work is necessary to improve and develop more effective techniques for industrial application.

1. Resin Solution

Although the resin treatment did enhance the fabric wrinkle resistant, the result is not much clear on the dimensional stability of cotton knitted fabric. In order to retain a higher effectiveness, a larger percentage of resin concentration should be added so as to obtain a more precise wrinkle resistant performance

2. Cover factor

Since the cover factor mainly affected the air permeability of the fabrics, further investigation on the cover factor can be examined in the future.

3. Wrinkle-free finishing

From the result that a higher concentration of Fixapret F-ECO finishing reduced the strength of knitted fabrics, other wrinkle-free finishes can be considered as a substitution in the experiment for enhancement of the properties of the fabrics.

4. Cotton-blend fabric

In this study, 100% cotton knitted fabrics were used for assessment. Other fiber blending of fabrics can be used to evaluate their wrinkle resistant, such as cotton/ polyester (80/20).

References

1. Abou-ana, M., Youssef, S., Pastore, C. & Gowayed, Y. (2003). "Assessing structural changes in knits during processing". Textile Research Journal, Vol. 3, Issue 6, pp.535-540.

2. Anand, S.C., Brown, K.S.M., Higgins, L.G., Holmes, D.A., Hall, M.E. & Conrad, D. (2002). "Effect of laundering on the dimensional stability and distortion of knitted fabrics". *Autex Research Journal*, Vol. 2, No2, pp.85-100.

3. Chippindale, P. (1963). "Wear, abrasion, and laundering of cotton fabrics". *Journal of the Textile Institute Transactions*, 1944-7027, Vol. 54, Issue 11, pp.T445 – T463.

4. Choi, K.F. & Lo, T.S. (2003). "An energy model of plain knitted fabrics". *Textile Research Journal*, Vol. 73, Issue 8, pp.739-748.

5. Ertugrul, S. & Ucar, N. (2000). "Predicting bursting strength of cotton plain knitted fabrics using intelligent techniques". *Textile Research Journal*, Vol. 70, Issue 10, pp.845-851.

6. Franck, E. & Reeves, W.A. (1962). "Some effects of the nature of cross-links on the properties of cotton fabrics". *Journal of the Textile Institute Proceedings*, Volume 53, Issue 1, pp.P22 – P36.

7. Gagliardi, D.D. & Wehner, A. (1967). "Influence of swelling and mowosubstitution on the strength of cross-linked cotton". *Textile Research*

Journal, Vol. 37, Issue 2, pp.118-128.

8. Hearle, J.W.S. (2007). "Physical structure and properties of cotton", University of Manchester, UK. In Cotton: Science and Technology, Cambridge: Woodhead, Chapter 1, pp.3-34.

9. Hsieh, Y.L. (2007). "Chemical structure and properties of cotton", University of California, USA. In Cotton: Science and Technology, Cambridge: Woodhead, Chapter 2, pp.35-67.

10. Hunter, L. (2007). "Testing cotton yarns and fabrics", CSIR and Nelson Mandela Metropolitan University, South Africa. In Cotton: Science and Technology, Cambridge: Woodhead, Chapter 12, pp.381-424

11. Knapton, J.J.F., Truter, E.V. & Aziz, A. K. M. A. (1975). "The geometry, dimensional properties, and stabilization of the cotton plain-jersey structure". *Journal of the Textile Institute*, Volume 66, Issue 12, pp. 413 – 419.

12. Kadolph, S. (2010). "Textiles". Upple Saddle River, NJ: Prentice Hall.

13. Lau, L., Fan, J., Siu, T. & Siu, L.Y.C. (2002). "Effects of repeated laundering on the performance of garments with wrinkle-free treatment". *Textile Research Journal*, Vol. 72, Issue 10, pp.931-937.

14. Lau, L.Y. (2001). "The performance of commercial wrinkle-free garments". *The Hong Kong Polytechnic University, Thesis for Master of Philosophy*.

15. Lo, L.Y. (2006). "Wrinkle-resistant finishes on cotton fabric using nanotechnology". *The Hong Kong Polytechnic University, Thesis for Doctor*

of Philosophy.

16. Lo, W.S., Lo, T.Y. & Choi, K.F. (2009). "The effect of resin finish on the dimensional stability of cotton knitted fabric". *Journal of the Textile Institute,* Vol. 100, Issue 6, pp.530 – 538.

17. Munden, D.L. (1959). "The geometry and dimensional properties of plain-knitted fabrics". *Journal of the Textile Institute,* Vol. 50, pp.T448-T464.

18. Mehta, S.B., Parikh, D.V. & Nanjundayya, C. (1967). "Resin-finishing *in situ*: the development of a process for imparting a crease-resistant finish to cotton fabrics". *Journal of the Textile Institute,* Vol. 58, Issue 7, pp.279 – 292.

19. Nutting, T. S. (1960). "Kinetic yarn friction in knitting". *Journal of the Textile Institute,* Vol. 51, pp.T190-T202.

20. Parker, J. (1998). "All about cotton: a fabric dictionary & swatchbook". Seattle, Wash.: Rain City Publishing.

21. Saville, B.P. (1999). "Dimensional stability". In Physical Testing of Textiles, Woodhead, Chapter 6, pp.168-183.

22. Ucar, N. (2007), "Cotton knitting technology", Istanbul Technical University, Turkey. In Cotton: Science and Technology, Cambridge: Woodhead, Chapter 9, pp.275-327.

23. Zhou, L.M. (2003). "Modification of ramie fabric with alkali and crosslinking treatment for wrinkle resistance". *The Hong Kong Polytechnic University, Thesis for Doctor of Philosophy.*

24. Wikipedia (2011), the free encyclopedia, "Cotton"

   http://en.wikipedia.org/wiki/Cotton (accessed on 9.4.2011)

25. Project Cotton (2008), The University of Missouri, "Cotton Classroom"

   http://cotton.missouri.edu/Classroom-Chemical%20Composition.html

   (accessed on 9.4.2011)

26. Cotton Incorporated (2011), "About Cotton"

   http://www.cottoninc.com/AboutCotton/How-Cotton-Is-Used/ (accessed on

   9.4.2011)

27. Bradberry, S. (2010), Knitting-and.com, "The Yarn Court"

   http://www.knitting-and.com/spinning/ycount.htm (accessed on 21.11.2010)